YOUR KNOWLEDGE HAS VALUE

AF 137200

- We will publish your bachelor's and master's thesis, essays and papers

- Your own eBook and book - sold worldwide in all relevant shops

- Earn money with each sale

Upload your text at www.GRIN.com and publish for free

An Introduction to Impedance Control of Constrained Robotic Mechanisms

Hayder Al-Shuka
Kareem Jawad Kadhim

Bibliographic information published by the German National Library:

The German National Library lists this publication in the National Bibliography; detailed bibliographic data are available on the Internet at http://dnb.dnb.de.

ISBN: 9783346307934
This book is also available as an ebook.

© GRIN Publishing GmbH
Nymphenburger Straße 86
80636 München

Print and binding: Books on Demand GmbH, Norderstedt, Germany
Printed on acid-free paper from responsible sources.

The present work has been carefully prepared. Nevertheless, authors and publishers do not incur liability for the correctness of information, notes, links and advice as well as any printing errors.

GRIN web shop: https://www.hausarbeiten.de/document/958860

An Introduction to impedance control of constrained robotic mechanisms

University of Baghdad, Department of Aeronautical Engineering, Baghdad, Iraq

Highlights
- Summarizing the categories of impedance control showing the features and the limitations of each category.
- Much attention is paid to variable impedance control considering the possible control schemes, performance and stability.
- Investigation of the limitations and features of multi-loop control for constant and variable impedance actuators.

Abstract
For a long time, the robotics community concentrated on improving the performance of the robotic systems in free space. The advanced control strategies for position control could not be sufficient to stabilize the motion of robots in constrained spaces. In general, there are two categories of force control schemes with miscellaneous subdivisions: hybrid position-force control and impedance control. The former is well suitable for well-known interaction environment, however, it does not consider the dynamic interaction between robot end-effector and the environment. In contrast, impedance control includes regulation and stabilization of robot motion via creating a mathematical relationship between the interaction forces and the reference trajectories. In general, a mass-spring-damper impedance filter is used for stabilization purposes. Tuning the parameters of the impedance filter is not trivial and may lead to unstable contact if an unsuitable strategy is used. The human, however, has amazing control systems with advanced biological actuators. He/she can manipulate the muscles stiffness to softly comply to the interaction forces. Accordingly, the parameters of the impedance filter should be time-varying rather than value-constant in order to meet the human behavior during interaction tasks. Accordingly, this paper is concerned with summarizing the categories of impedance control showing the feature and the limitations of each category. Much attention is paid to variable impedance control considering the possible control schemes, performance, the stability, the integration of constant/variable compliant elements with the host robots, etc.

Keywords
Impedance control, Force control, Admittance control, Variable impedance control, Series elastic actuators, Variable impedance actuators

2

1. Introduction

When a robot is in contact with the environment via its end-effector, some important points should be noted:

- Position control strategies are not sufficient for precise tracking of the desired position and force references; both the position and interaction force should be controlled carefully. For example, if the task of target robot is to write something, neglecting control of interaction force may lead to either loss of contact or hard pressure on the target environment [1]. In general, for rigid or dynamic interaction environments, pure position control schemes are not recommended especially if the environment is stiff; the contact forces may reach unsafe values [2].

- Besides, the robot looses some degrees of freedom (DoFs) during the contact phase. Consequently, the generalized coordinates of the target robot could be larger than its DoFs due to its constrained motion; it constitutes closed chain mechanism with redundant coordinates [3].

- The robot may change its configuration during a transition from open chain mechanism to a closed chain. In effect, three motion phases could be produced: free motion phase, contact motion phase (impact phase), and constrained motion phase. Every phase may have its own features and control law [3].

One of the solutions to regulate and control the interaction forces is hybrid position/force control coined by Raibert and Craig [4]. The hybrid force-position control decouples the task space into position-controlled space and force-controlled space. Then the hybrid position/force control law is designed for tracking the desired position and force references. However, this scheme neglects the impedance effect between the environment and the robot end-effector.

In effect, the impedance control plays an important role in any workspace that involve human-robot interactions. The human has the amazing adaptability to change the muscle impedance (e.g. stiffness) when contacting with unknown environments. If the environment is stiff, the robot should be soft and vice versa. However, robots not have this feasibility; they are stiff in principle. They are well suited for precise free-motion space, but unstable problems could occur when moving in unstructured environments. The excessive interaction forces should be avoided. This can be performed by making the robots changing their stiffness. Therefore, Hogan has proposed active impedance control which is inspired from the biomechanics of human motion in free and constrained spaces [5,6]. The idea behind impedance control is to design a user-defined dynamic relationship between the reference trajectory of the end-effector and the interaction contact force/torque along each axis. However, a trade-off between the tracking of the position and the interaction forces occurs. The position controlled axes have large stiffness while the force controlled axes have small stiffness [7]. Hogan proposed two models of impedance control [5,6]: torque or force-based impedance control, and position-based impedance control. Due to the associated limitations of conventional impedance control, Anderson and Spong [8] have proposed the hybrid impedance control. The idea is to exploit the concept of hybrid position/force control and integrate it with the impedance control. A robust control version of hybrid impedance control with inner acceleration is proposed by Liu and Goldenberg [9] such that the convergence of position, velocity, and acceleration are proved with bounded force error.

Zhu and Schutter [10] have made a link between the hybrid control and impedance control using the virtual decomposition control for a two six-joint industrial robots. Extensions of other schemes were proposed for force-tracking impedance control [11-18]. The force tracking impedance control motivates the researchers towards variable impedance control which means that the target impedance parameters change adaptively for safe interaction motion with an unknown environment. This behavior is similar to the human 's one which has adaptable flexibility and impedance to deal with interaction tasks. The challenge of variable impedance control may lie in how the analyst can select the impedance time-varying parameters with guaranteed stability. The same problem may be encountered when dealing with fixed impedance parameters [7,19]. The limitations inherent in active impedance control lead to creating new generations of actuators technology started from series elastic actuators to variable impedance actuators [20]. These actuators include constant or variable compliant elements in their design [21,22]. The idea behind these actuators is to imitate the behavior of human motion during contact with unknown constrained spaces. By controlling the stiffness of the target robot, the robot can adaptively comply with the interactions forces and generate safe contact. How to integrate the active impedance control with these passive design-based actuators is not trivial and encounters some control and stability problems.

Remark 1: In general, there are three possible compliance schemes for making the robotic system compliant:

- Pure passive compliance, e.g. the remote center of compliance used in industrial robot applications without force feedback.
- Active compliance control, e.g. hybrid force-position control and impedance control schemes, etc.
- Active compliance control integrated with passive elements, e.g. design and control of series elastic actuators (SEA).

For more details on features and drawbacks of these compliance classifications, please see [23,2,24].

Remark 2: The active compliance control can be classified as direct force control, e.g. explicit force control and hybrid force/position control schemes, and indirect force control schemes represented by the impedance control. For more details, the reader is referred to [23,2,24].

In light of above, this paper is concerned with summarizing the different schemes of impedances control with miscellaneous control modes. The features and disadvantages of each scheme are introduced with some details. The stability problems and applications of the impedance control are also presented. Accordingly, the paper can be organized as follows. Section 2 introduces background about the two main modes of conventional impedance control. The force tracking position-based impedance control is investigated in Section 3. Whereas, Section 4 discusses the scheme of variable impedance control. The impedance control of flexible joints with constant and variable compliance is presented in Section 5. Section 6 concludes.

2. Background

Although the impedance control schemes are referred to as indirect force method, some of these schemes can include force tracking loop with the impedance target, e.g. the position-based impedance control (admittance control) can be modified to improve the interaction force tracking problems; this will be discussed in some details in the next sections. As aforementioned, the idea of the impedance or admittance control is to generate a dynamic relationship between the interaction force/torque and the position trajectory of the robot end-effector by using the virtual mass-spring-damper system, please see Fig. 1 for a description of the idea of impedance control.

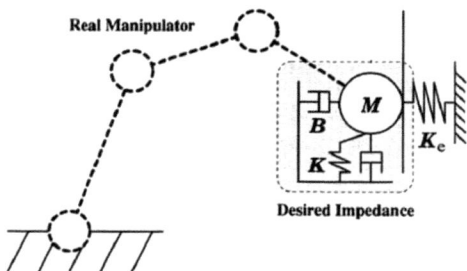

Fig. 1 The description of impedance control for robot in contact with external environment [26]

By tuning the parameters of the impedance parameters, a suitable performance can be obtained for the host robot; there is a deviation in robot motion associated and coupled with the deviation of interaction force. Basically, impedance control may consist of two nested control loops: an outer impedance control loop and an inner position or force control loop. For more details on the interaction force control schemes, please see [23,25]. Below we will describe modeling of robots with interaction forces and how to describe the suitable formulation of impedance control for redundant and non-redundant robots. Then the two conventional categories of impedance control are described with some details.

Remark 3: There are three related sub-categories of impedance control which are: stiffness control [1,], compliance control [24], and damping control [27]. For more details, the reader is referred to the mentioned references.

2.1 Dynamic modeling of robots in constrained motion

The target impedance dynamics (outer impedance loop) is preferred to be expressed in terms of the task coordinate frames because the task geometry may decide which directions are motion-constrained and force-sensitive [7]. The task specifications for motion and interaction forces and the force feedback are closely related to the end-effector. Description of the dynamic behavior of the end-effector and its association with the external environment is essential for high-performance manipulator control [28]. In general, the impedance control consists of two control loops: an outer impedance loop regulating the

5

interaction between the end-effector and the external environment, and an inner control loop which could be torque control loop or position control loop. For outer impedance loop, representation of the dynamics of the impedance target in terms of task space is necessary as mentioned previously. For the inner control loop, there are two possibilities of coordinate representation for the control law. For the force-based impedance control, the inner joint space torque control requires a transformation of the commanded forces generated by the outer loop into commanded torque that should be tracked Accordingly, it requires the calculation of the Jacobian online. For the position-based impedance control (admittance control), the inner position control can be represented in joint space by transformation of the commanded task coordinates into joint coordinates using inverse kinematics [29-31]. However, the inner position control law can be represented in task coordinates as made in [32]. In effect, despite the usefulness of task space formulation for implementation of high-performance control schemes, the measurements of end-effector position and orientation is not trivial; vision technology may be necessary for this purpose. On the other hand, implementation of joint space control combined with Cartesian impedance outer loop may require calculation of Jacobians, forward and inverse kinematics schemes which could be computationally complex.

Consider $n - DoF$ robot operating in m-dimensional Cartesian coordinates. The kinematics of the host robot can be expressed as

$$x = T(q) \tag{1}$$
$$\dot{x} = J(q)\dot{q}, J(q) = \frac{\partial k(q)}{\partial q^T} \tag{2}$$
$$\ddot{x} = J(q)\ddot{q} + \dot{J}(q, \dot{q})\dot{q} \tag{3}$$

where $x \in \mathbb{R}^m$ is the Cartesian position of the end-effector, $T(.): \mathbb{R}^n \rightarrow \mathbb{R}^m$ denotes to the forward kinematics, $q \in \mathbb{R}^n$ is the joint position, and $J(.) \in \mathbb{R}^{m \times n}$ is the manipulator Jacobian matrix. For robots in constrained space, the 2nd Lagrangian formulation can be used for modelling. So, the dynamic equation of motion for constrained motion can be expressed in joint space as follows.

$$M(q)\ddot{q} + C(q, \dot{q})\dot{q} + g(q) = J(q)^T f_e + \tau \tag{4}$$

where $M(.) \in \mathbb{R}^{n \times n}$ represents the inertia matrix, $C(.) \in \mathbb{R}^{n \times n}$ is the Coriolis and centripetal matrix, $g(.) \in \mathbb{R}^n$ is the gravity, $f_e \in \mathbb{R}^m$ denotes to the external and interactional force, and $\tau \in \mathbb{R}^n$ is the input control torque.

For a robot having $(m = n)$, i.e. the number of the generalized coordinates is equal to the task space coordinates, the robot is non-redundant. Whereas, if $n > m$, i.e. the number of the generalized coordinates is more than task space coordinates, the robot is called kinematically redundant robot with redundant coordinate $r = n - m$.

Remark 4: Equation (4) can be transformed into task space coordinates by using the kinematic relationships of Eqs. (2) and (3). For task-space dynamics of non-redundant and redundant robots, the reader is referred to [28,29] respectively.

In general, there are two possible aspects of redundancy problems which are motion redundancy and torque redundancy [26,33,34]. For better performance of redundant robots, the null space dynamics should be considered, dealing with only task space

6

impedance control may not be sufficient. For more details on redundant robots and the modified impedance control, the reader is referred to [35-37].

2.2 Force or torque-based impedance control

The idea behind the force-based impedance control (it is simply called impedance control in the literature) is to make the controller react to the motion deviation by generating forces [2,29]; it consists of two control loops: outer position loop and inner force loop. The controller may attempt to stiffen a soft force source [29], please see Fig. 2 for a generic description of force (torque)-based impedance control.

To motivate the concept of the impedance control, consider the following simple second order system, see Fig. 3

$$m\ddot{x} + b\dot{x} + kx = u + f_e \tag{5}$$

where x and the first two derivatives represent the state variables of the system, m is the mass of the investigated system, b represents the damping coefficients, k denotes to the system stiffness, u is the input control, and f_e is the external force affecting the system (it could be the interaction contact force or any external force).

As stated previously, the impedance control attempts to make a dynamic relationship between the interaction force and position error by assuming virtual mass-spring-damper model with desired trajectory; accordingly, the target impedance function can be expressed as [1]

$$m_i(\ddot{x} - \ddot{x}_r) + b_i(\dot{x} - \dot{x}_r) + k_i(x - x_r) = f_e \tag{6}$$

where m_i, b_i, and k_i are the target impedance coefficients that govern the performance of the controller. x_r represents the desired equilibrium trajectory that should be slightly inside the contact environment to keep contact.

Changing the structure of the target impedance dynamics or the behaviour of the target impedance coefficients leads to different impedance control strategies. Anyway, taking Laplace transformation for eq. (6), the following equation can be obtained

$$sZ(s) = m_is^2 + b_is + k_i \tag{7}$$

with

$$sZ(s) = \frac{F_e(s)}{E_e(s)} \tag{8}$$

where $E_e(s)$ is the position error due to the interaction forces, and $Z(s)$ represents the mechanical impedance of the host system. Substituting eq. (6) into eq. (5) may lead to the following closed loop control system

$$u = (b - mm_i^{-1}b_i)\dot{x} + (k - mm_i^{-1}k_i)x - (1 - mm_i^{-1})f_e + mm_i^{-1}(b_i\dot{x}_r + k_ix_r) + m\ddot{x}_r \tag{9}$$

As we see the feedback controller of eq.(9) needs the measurements of interaction force and the state variables of the end-effector.

Let's consider the case $m = m_i$, then eq. (9) is simplified to

$$u = (b - b_i)\dot{x} + (k - k_i)x + (b_i\dot{x}_r + k_ix_r) + m\ddot{x}_r \qquad (10)$$

which represents classical velocity and position feedback control with feed-forward term denoted by the desired acceleration.

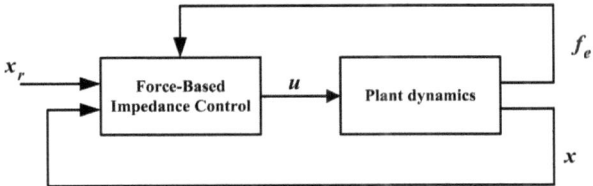

Fig. 2: The generic schematic diagram of force-based impedance control [30,31].

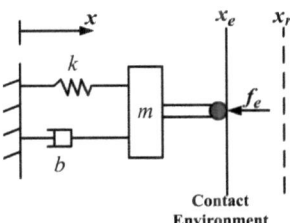

Fig. 3: The systems dynamics in contact with external environment. The coordinate x_e represents the environment position [49].

Remark 5: In effect, there are three possible models for representing the target impedance dynamics that correlates the dynamic relationship between the position and contact forces

$$I. m_i(\ddot{x} - \ddot{x}_r) + b_i(\dot{x} - \dot{x}_r) + k_i(x - x_r) = f_e \qquad (11a)$$
$$II. m_i(\ddot{x}) + b_i(\dot{x} - \dot{x}_r) + k_i(x - x_r) = f_e \qquad (11b)$$
$$III. m_i(\ddot{x}) + b_i(\dot{x}) + k_i(x - x_r) = f_e \qquad (11c)$$

Seraji and Colbaugh [38] have used both eqs. (11a,c) to derive the equations of the steady-state force and position errors respectively, whereas, Yoshikawa [1] has used eq. (11b) to derive the impedance control for both free and constrained spaces. He proved that when there is no contact force, the controller represents position and velocity feedback control. Huang and Chien [32] and Deluca [39] have used the feedback linearization-based impedance control for controlling robots in contact with the rigid environment. They have used eq. (11a) as the target impedance dynamics. Khan et al [40]. have used adaptive impedance control based on the target impedance dynamics of eq. (9a) for upper limb assist exoskeleton. In general, the following important points should be noted [7,39,41]:

- These target dynamics models can be changed according to the purpose of the control problem. For example, force tracking-based impedance control may require modifying the right-hand side of eq.(11) such that it may include force error rather than environment force only, for details please see the next section.
- Eq. (11) makes a compromise between the position and the contact forces such that there could be deviations in the desired position and the force references.
- The variable x_r represents the desired impedance trajectory which should be tracked by the inner position control. In effect, this impedance variable should be slightly inside to keep contact.
- Eq. (11b) could mean that the desired acceleration variable is zero while eq. (11c) could represent target dynamics with zero velocity and acceleration impedance trajectory.
- Since a control loop based on force error is missing, forces are only indirectly assigned by controlling position.
- The choice of a specific stiffness in the impedance model along a Cartesian direction results in a trade-off between contact forces and position accuracy in that direction.
- It is preferable to mimic the behavior of motion of human arm such that the motion is fast and stiff in free space and slow and compliant in constrained space.
- Large m_i and small k_i in Cartesian coordinates with contact space may lead to the low contact force.
- Large k_i and small b_i in Cartesian coordinates with free space may lead to good motion tracking.
- The advantage of damping coefficient b_i is to shape the transient response.
- As a rule of thumb, the stiffer the environment, the softer is the impedance stiffness k_i.
- The external environment force can be eliminated by substituting eq. (6) into eq. (5) such that the contact force is neglected, however, the measurement of acceleration is needed in this case which is very noisy.

2.2 Position-based impedance control (Admittance control)

In admittance control, the controller attempts to soften the stiff position source via reacting to the interaction forces by imposing deviation from the desired motion 2,29]. Position-based impedance control consists of two control loops: inner position loop for controlling the compliant position references and an outer loop for providing the desired target impedance dynamics that delivers the desired compliant references; please see Fig. 4 for general description of the position based impedance control.

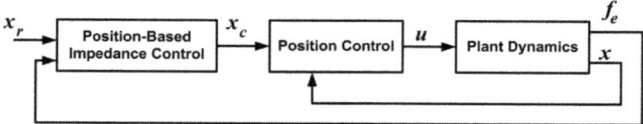

Fig. 4 The schematic diagram of position-based impedance control [30,31]

To understand the notion of the position-based impedance control, consider the following 2nd order dynamic system

9

$$m\ddot{x} + b\dot{x} + kx = u + f_e \tag{12}$$

with nomenclatures mentioned previously.

Eq.(9) should be modified to isolate the inner position control loop from the outer impedance control loop. This can be made by introducing a new variable called impedance reference trajectory, x_i, for the end-effector resulted from the desired references of the end-effector and the measurement of the interaction force wrench, please see Fig. 5 for some details. Accordingly, the outer impedance filter can be expressed as [42-47]

$$I. \, m_i(\ddot{x}_c - \ddot{x}_r) + b_i(\dot{x}_c - \dot{x}_r) + k_i(x_c - x_r) = f_e \tag{13a}$$
$$II. \, m_i(\ddot{x}_c) + b_i(\dot{x}_c - \dot{x}_r) + k_i(x_c - x_r) = f_e \tag{13b}$$
$$III. \, m_i(\ddot{x}_c) + b_i(\dot{x}_c) + k_i(x_c - x_r) = f_e \tag{13c}$$

The inner position control can be implemented using PID family in our simple example, so the control law can be expressed as

$$u = k_p(x_r - x) - k_v \dot{x} \tag{14}$$

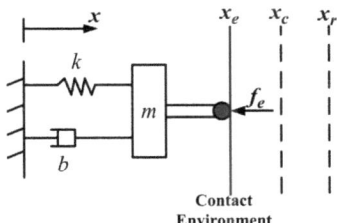

Fig. 5 The systems dynamics in contact with external environment. The philosophy of impedance target dynamics changes due to adding the virtual impedance reference trajectory x_c [49]

In effect, the well-known nonlinear schemes, e.g. feedback linearization control (computed torque control), passivity-based control, robust sliding mode control, mode reference adaptive control, etc. can be used for the inner position control loops [42-48, 32].

2.3 Position-based impedance control vs. force-based impedance control

In effect, the force-based impedance control and the position-based impedance control are based on assumption of force-controlled system and position-controlled system; therefore, their performance and stabilities could be different [41]. There are some important points that should be considered when using impedance control [41,29, 26]:

- Since most of the industrial electro-mechanical manipulators are equipped with servo position control loops, the position-based impedance control could avoid redesigning the inner position loop.

- For desired stiff impedance behaviour, the force-based impedance control may encounter instability problems due to the amplification of noise. It is known that if the environment is soft (compliant), the stiffness of the end-effector should be stiffer and vice versa. Accordingly, a force-based impedance control could be suitable for interaction with a stiff environment. In contrast, the position-based impedance control is more suitable to implement stiff behaviour than compliant behaviour; it is suitable for interaction with compliant environment.
- The performance and the stability of force-based impedance control may depend on back drivability and the amount of friction for the host system, whereas, the performance of the position- based impedance control may depend on the performance of the inner position control.

For more details on the differences between these categories of impedance control, the reader is referred to [41].

4. Position-based impedance control with force tracking

The stiffness of muscles plays an important role in dextrous and robust motion for human beings. For example, the human arm can control the interaction contact force by modifying its muscle's stiffness such that the interaction contact force can be either increased by making the arm stiffer or decreased by reducing the arm stiffness. In addition, he/she can keep the force tracking error within a specified range in the presence of disturbances and uncertainty [49]. In effect, the target impedance dynamics of eqs. (11) and (13) is asymptotically stable in free space while there are steady state position and force errors in constrained space. Accordingly, the exact position and force tracking may not occur in impedance control strategies. Please see Fig. 5 for large view of the philosophy of robot-environment interaction. The main limitation of impedance control is that the interaction forces are controlled indirectly by selecting the desired impedance dynamics. However, this may demand accurate knowledge for environment parameters (e.g., environment location and stiffness) which are difficult to be specified in practical applications [38,50].

4.1 Dependence of force tracking error on the knowledge of the environment parameters

To see the importance of knowledge of environment parameters, consider the following scalar target impedance function

$$m_i(\ddot{x}_r - \ddot{x}_c) + b_i(\dot{x}_r - \dot{x}_c) + k_i(x_r - x_c) = f_r - f_e \qquad (15)$$

replacing the environment contact force by the difference between the required desired force and the sensed contact force. This modification could be necessary for force tracking; other modification in impedance function will be clarified later.

If the desired reference trajectory maintains constant values such that their first and second derivatives are equal to zero. So, eq. (15) becomes

$$-m_i\ddot{x}_c - b_i\dot{x}_c + k_i(x_r - x_c) = f_r - f_e \qquad (16)$$

Using a simple spring model for representing the deformation of environment (the environment stiffness dominates its deformation), the interaction force can be expressed as

$$f_e = k_e(x - x_e) \tag{17}$$

Rewriting the above equation to get the end-effector position

$$x = \frac{f_e}{k_e} + x_e \tag{18}$$

Inserting the force error in eq. (18) leads to

$$x = \frac{(f_r - e_f)}{k_e} + x_e \tag{19}$$

with $e_f = (f_r - f_e)$

Because the objective of the inner position control loop is to track the commanded compliant impedance references (x_c), a position error could be produced. Accordingly, x_i can be expressed as

$$x_i = \frac{(f_r - e_f)}{k_e} + x_e + e_p \tag{20}$$

with

$$e_p = x_c - x = x_c - \frac{(f_r - e_f)}{k_e} - x_e \tag{21}$$

Substituting eq.(20) in eq. (16) produces the following force/position error differential equation (closed loop)

$$m_i \ddot{e}_f + b_i \dot{e}_f + (k_i + k_e)e_f = k_e k_i x_e + m_i \ddot{f}_r + b_i \dot{f}_r + k_i f_r + k_e(m_i \ddot{e}_p + b_i \dot{e}_p + k_i(e_p - x_r) \tag{22}$$

If the impedance system reaches the steady state region assuming the desired environment force is of constant value, the steady state force error can be expressed as

$$e_f = \frac{k_i k_e}{k_i + k_e}\left[\frac{f_r}{k_e} + x_e + e_p - x_r\right] \tag{23}$$

Let $x_r = x_e + \frac{f_r}{k_e}$ which *includes good knowledge of environment parameters*, eq. (23) reduces to the following equation

$$e_f = \frac{k_i k_e}{k_i + k_e} e_p \tag{24}$$

in which *the position error plays an important role in steady state force error.*

4.2 Methods

In general, most robotic systems may require being in contact with the external environment. Regulation of the interaction force is necessary to avoid instability and unsafety problems. Some robot applications include controlling and stabilization of constant value interaction force, such as with deburring, welding, grinding, etc [51]. Human-robot interaction applications, nonetheless, demand time-variant interaction forces, such as robot-aided cell injection [52,121] and rehabilitation applications [54,55]. Accordingly, conventional impedance control could not be suitable to these application and large deviations of position and forces could be produced; tracking time-varying force control combined with impedance behavior is demanded.

According to the previous discussions and eq. (22), it has been shown that convergence of interaction force tracking could not be ensured in position-based impedance control especially with uncertain environment stiffness and uncertain modeling of the host robotic system [56]. Literature proves that there are two distinguished techniques that attenuate the force tracking error which are:

- modification of the reference trajectory combined with estimation of environment geometry and physics, and
- modification of the target stiffness which could control the required interaction force carefully. In effect, making the system stiffness variable to imitate the human behavior is considered as variable impedance control which will be explained extensively in the next section.

For more details on this interesting topic, please see [50, 11-18]. Table 1 shows some important schemes for force tracking with some detailed description.

Table 1: Position-based impedance control with some schemes for tracking interaction force references

Ref.	The dynamics of impedance behaviour (the outer impedance control loop)	The modified controlled variable	Comments
[56]	$m_i(\ddot{x}_r - \ddot{x}_c) + b_i(\dot{x}_r - \dot{x}_c) + k_i(x_r - x_c) = f_e$ (25) with $k_i = \dfrac{\left(K_p e_f + K_d \dot{e}_f + K_i \int_0^t e_f(\tau)d(\tau)\right)}{(x_r - x_i)} + k_0$ $x_r = x_r^0 - \gamma(x_r - x_c)$ (26) where k_i represents the variable impedance stiffness, K_p, K_d and K_i denote to the feedback gains, k_0 denotes to initial impedance stiffness, x_r^0 represents the initial desired equilibrium trajectory, and γ is constant.	The target stiffness and the desired reference trajectory	1. The desired reference trajectory is modified due to eq. (26) to compensate for the force error. 2. The commanded compliant trajectory would be slightly penetrated into the interaction environment in order to keep contact. 3. The target impedance stiffness is variable and represents PID controller in terms of contact force error and the model-following error. 4. The proposed impedance control scheme enables the manipulator to track the reference force in the steady state despite a sudden change in the trajectory of the environment.
[49]	Impedance function: $m_i(\ddot{x}_r - \ddot{x}_c) + b_i(\dot{x}_r - \dot{x}_c) + k_i(t)(x_r - x_c) = f_e$ (27) with $k_i(t) = (k_p e_f + k_d \dot{e}_f)/(x_r - x_c)$ (28) Please see [49] for more detailed on the notations used.	The target impedance stiffness	The authors preferred to change the target stiffness because the modification of the desired reference trajectory is unintuitive and the small change of x_r may lead to drastic changes in environment forces. 2. The target stiffness is variable and represents PD controller of contact force error. Its value could be negative or time varying according to the proposition of the authors.
[38]	The authors suggested two control schemes as follows. (i) Controller 1 (MRAC) The target impedance function: $m_i(\ddot{x}_r) + b_i(\dot{x}_r) + k_i(x - x_r) = f_r - f_e$ (29) with MRAC controller for the reference trajectory $x_r = g(t) + k_p e + k_v \dot{e}$ (30)	The desired reference trajectory	For controller 1: 1. The derivative of force error needs calculation of derivative of contact force which is undesirable. There are two possible techniques for solving this problem: (1) making a filter for the sensed force signal then differentiate the filtered signal [57], or (2) by exploiting the simple spring model for the

with $g(t)$ representing auxiliary signal.

(ii) Controller 2 (indirect adaptive control):
The target impedance function:

$$m_i(\ddot{x}_r - \ddot{x}_c) + b_i(\dot{x}_r - \dot{x}_c) + k_i(x_r - x_c) = f_r - f_e \quad (31)$$

with

$$x_r = \hat{x}_e + \frac{1}{\hat{k}_e} f_r \quad (32)$$

$$\dot{\hat{k}}_e = -\gamma_1 x(\hat{f} - f) \quad (33)$$

$$\dot{\hat{x}}_e = \frac{(\hat{f}-f)}{\hat{k}_e}(\gamma_1 x \hat{x}_e + \gamma_2) \quad (34)$$

where the symbol $(\hat{\cdot})$ represents the estimated value, γ_1 and γ_2 are constants.

environment [58].
2. The reference trajectory for the end-effector is modified to compensate for the contact force error. According to Eq (30), the reference trajectory is expressed in terms of the states of the force error using the MRAC.
3. This algorithm does not need the environment parameters such that it could be robust to environment changes related to geometry or properties.

For controller 2:
1. The unknown parameters of the contact environment are estimated based on Lyapunov's stability.
2. The desired reference trajectory of the end-effector is calculated based on the estimated values of the environment parameters which can guarantee stability and good tracking for contact force.

In summary, the following points should be considered:

- The impedance force control should be able to achieve three important aspects: (1) minimization of position error due to the modelling uncertainty, (2) the desired interaction force references should be tracked carefully with guaranteed stability, and (3) the proposed controller should be robust to uncertain position and stiffness of the environment [16].
- Most researchers assume decoupled outer impedance filter which simplifies the control problem; the stability analysis and performance of the proposed force tracking impedance filter may depend on linear control theory such root locus, Hurwitz 's stability, etc.
- One important observation is that the virtual stiffness of the impedance behaviour could lead to some steady state errors; therefore, cancelling it may lead to zero steady state error [16, 59].
- One of the interesting key point for tracking the interaction force is to generate desired variable stiffness for force tracking purposes. This philosophy is closer to the interaction motion control of human.
- The variable impedance behavior can be combined with force tracking schemes to perform high-level control strategies, please see the next section.

5. Variable impedance control

For most biological movements, muscles behave as mechanical actuators with a nonlinear stiffness behavior; the muscle viscosity could be considered constant according to biological studies. The force-velocity relationship includes nonlinear characteristics during contraction and stretching; increasing the applied force may result in an increase in muscle stiffness. One important thing that should be noted is that the slopes of the force-velocity curve represent the muscle impedance associated with muscles movements [60] and the references therein. The interaction of human muscles with external interaction forces can be represented by two interacting subsystems as shown in Fig. 6.

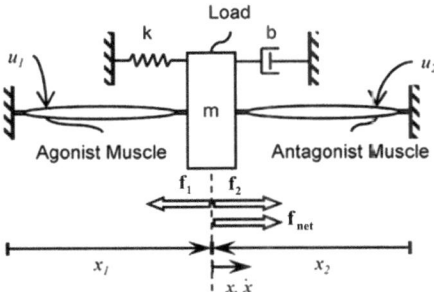

Fig. 6 A simple schematic for muscle-load interaction. x_1, x_2 are the lengths of the muscles and u_1, u_2 are the input controls, for more details see [61].

One subsystem represents the modeling of the antagonistic muscles and the second system represents the generalized forces; the generalized forces are represented by a virtual spring-mass-damper with constant/variable coefficients. This modeling meets active impedance control of robots described previously. Besides it is well-known that locomotion of humans consists of miscellaneous motion phases, e.g. single support phase, double support phase, jumping, etc. So, the human should modify the muscle stiffness to attenuate any heterogeneous disturbances or even to track some desired interaction forces; the human control system is provided with variable stiffness characteristics [62]. Using the conventional impedance control with fixed coefficients, e.g. fixed stiffness, cannot achieve the required target impedance. In addition, the human can grasp the objects softly and safely by regulating the muscles stiffness. Accordingly, variable stiffness-based impedance control can improve the performance of the desired force tracking and the dexterity of the robotic system. This policy of changing stiffness is explained in the last subsection and revisited here for its correlation of variable impedance control, please see the related refs. [49,59,63].

The idea behind variable impedance control is to propose a suitable modulation strategy for the parameters of the impedance behaviour such that the stability is guaranteed and the performance is better and safer. Mathematically, the target impedance behaviour with variable parameters can be expressed as:

$$m_i(t)(\ddot{x}_r - \ddot{x}_i) + b_i(t)(\dot{x}_r - \dot{x}_i) + k_i(t)(x_r - x_i) = -f_e \tag{34}$$

with possibly time-varying $m_i(t)$, $b_i(t)$, and $k_i(t)$.

In light of above, there are two main objectives for equipping the target impedance with variable stiffness:

- for tracking of interaction force references.
- for increasing the adaptability, imitation of the biological behaviour during contact with different environment stiffness, and guaranteeing the coupled stability of the robot-environment interaction.

Ferraguti et al. [64] have proposed a reservoir (tank)-based impedance control with variable stiffness matrix to imitate the surgeon during the punctuating tasks. The authors investigated some important issues related to stability and performance of outer impedance filter with nonlinear behavior. They augmented the impedance model with an energy storing element whose role is to store the energy dissipated by the controlled system. With this scheme, the impedance control with variable mass and stiffness matrices can be a powerful tool to deal with a compliant environment which requires time-varying interaction forces. Besides, the following should be noted:

- The analyst can avoid the need for measurement of interaction forces if the desired virtual impedance has a variable mass matrix of the host robots with constant desired damping and stiffness matrices.
- The variable stiffness impedance behavior associated with inertia variable matrix may lead to unstable behavior violating the passivity condition; hence, the tank-based impedance control was proposed to solve this problem.

Changing the stiffness matrix of the target impedance dynamics of the investigated system could increase the robustness of the system toward uncertain parameters [65]. This is motivated by the idea that the stiffness matrix of the target impedance is the dominant term for steady state region and low-frequency interaction forces. If the applied contact forces at the tip position of the end-effector are too large, this may excite the disturbances resulting from friction, measurement noises, and the unknown profile of contact surface [66]. Accordingly, the robotic system should adapt its impedance to compensate for these disturbances. In light of above, Yuan et al. [67] have proposed variable impedance control for lower limb rehabilitation device. The idea is to design the target impedance dynamics for the impaired limb such that the desired stiffness of the limb imitates the sound human stiffness during extension and flexion. The electromyogram (EMG) technique is used for transferring the desired variable stiffness to the rehabilitation device. The proposed impedance behavior was based on two determinants: knee damping and stiffness. A constant-value damping parameter is used according to the author's experiments; please see Fig. 7 for a description of a control system for lower limb rehabilitation.

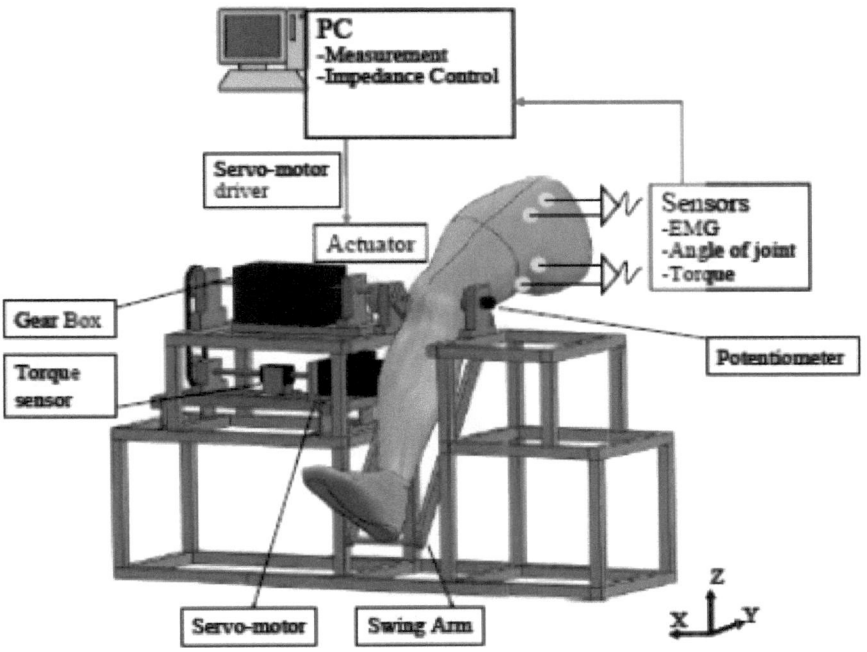

Fig. 7 Diagram for rehabilitation robotic system for lower limb [67]

Park and Cho [65] have proposed varying stiffness-based impedance control for parallel robots without measurement of tip position. The idea is to change the system stiffness in order to compensate for the model error. Ollinger et al. [68] have proposed active impedance control for lower limb exoskeletons in order to adjust the energy transfer between the active exoskeleton and the impaired limb. It is an alternative tool to EMG-based control for estimation of muscle torques which is a rather complicated approach. Tsumugiwa et al. [69] have proposed variable impedance control for human-robot cooperation. The idea is to adjust the target damping coefficient of the robot impedance function proportional to the estimation of the arm stiffness of the human operator. This procedure may avoid instability due to the increased stiffness of the operator's arm. Even if the stiffness of the human operator's hand is very high, introducing a low damping coefficient for the target impedance of the robot may lead to stable operation.

Ficuciello et al. [59] have improved the performance of impedance control of a 7-degree-of-freedom KUKA LWR4 by exploiting the kinematic redundancy and modulation of impedance parameters such that they imitate the human behavior. The authors have found that the redundancy may enlarge the stability margins of the impedance parameters. In addition, the virtual variable impedance behavior with convenient modulation of the time-varying parameters could be superior compared to constant coefficients impedance behavior. The variable impedance target may: (1) enhance the performance and the safety of interaction tasks with the human, and (2) compromise between the accuracy and the execution time.

In summary, the following important points should be noted during the design of active variable impedance control for human-robot interaction:

- There are two possible strategies for imitation of human impedance behavior: the variable impedance actuators [70-73,20], and the active impedance control with suitable modulation for the time-varying tuned parameters.

- The challenge of application of variable impedance target with human behavior is how to transfer the impedance characteristics from the human to the robot. In effect, there are some possible techniques used for estimation of human impedance where most of them are based on neurological schemes, such as the human central nervous system [74], learning control strategy [75], and tele-impedance [76].

- In general, if the robot should be freely driven by the human, the robot impedance should be low; zero stiffness is recommended in this case. For fast motion purposes, the virtual robot damping should be decreased and vice versa. Whereas, decreasing the virtual inertia may lead to instability problems [59].

- On the other hand, the regulation of the virtual stiffness of the host robot is necessary for surgery and rehabilitation applications and collaborative robots [59,77]. Most of the researchers have concentrated on the manipulation of the system's stiffness due to the following points: (1) the wide range of adjustability of the system stiffness compared with damping and inertia coefficients, (2) the stiffness term is the dominant factor for low-velocity motion, and (3) the stiffness term has large effect on the system stability for the steady state regions. For more details on characteristics and details of human arms and legs stiffness, the reader is referred to [78-80,81].

- With constant-parameters-based impedance control, the passivity property is conserved, while with arbitrary time-varying parameters for the impedance behavior, the passivity property could be lost [19]. Although the strength strategy of tank-based impedance control for generating stable interaction forces with variable stiffness [64, 82], it is dependent on the states of the system which means that it should be applied online. Kronander and Billard [19] have proposed states-independent scheme to ensure the stability of variable impedance control; this means that the time-varying parameters of the target impedance behavior could be applied offline before execution of the task. Please see Fig. 8 for some details of the proposed control system.

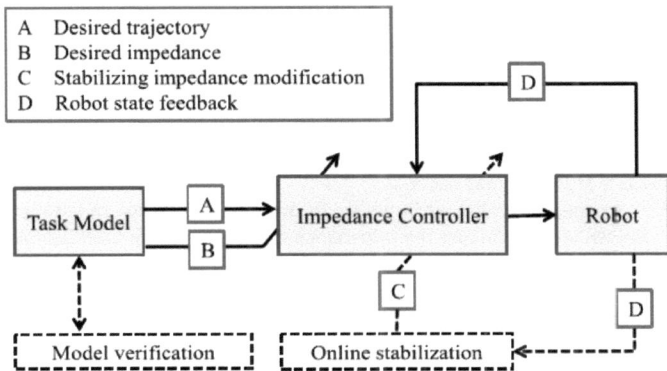

Fig. 8 The proposed strategy for variable impedance control with variable tuned parameters [19]

6. Active impedance control flexible joints actuated systems with inner torque loop

Cascade control is used widely in process applications with different time-scale subsystems. In a single-input-single-output (SISO) control loop, the set-point reference is the input to the controller while its output drives the plant. For example, the position controller driving the torque of the DC motor to keep the position at its set point. In a cascade control arrangement, there are two (or more) controllers of which one controller's output drives the set point of another controller, e.g. the multi-loops control of flexible jo nt where three possible loops could be nested such as position, torque and velocity loops The controller driving the set point (the position controller in the example above) is called the primary, outer, or master controller. The controller receiving the set point (torque controller in the example) is called the secondary, inner or slave controller. Figure 9 shows a schematic diagram for two nested cascade control.

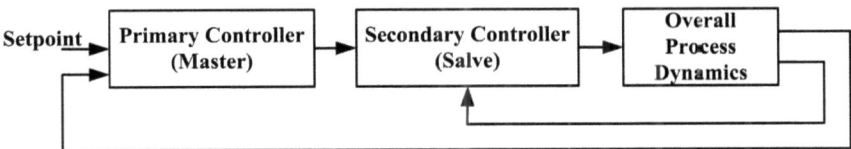

Fig. 9 The generic block diagram of two nested cascade control

In general, there are some requirements which should be considered to guarantee better performance for cascade control [83]:
- The inner loop has influence on the outer loop.
- The dynamics of the inner loop are faster than those of the outer loop.
- The inner loop disturbances are less severe than the outer loop disturbances.

On the other hand, some challenges should be noted while designing the cascade control [83]:

- The overall equipment costs due to the additional sensors and controllers.
- Cascade control systems are also more complex than single-loop controllers, requiring twice as much tuning.
- The inner control loop should be three times faster than the outer loop, otherwise instability problems may occur.

For more comprehension on the performance of the cascade control, the reader is referred to [84, 85]. Below, we will discuss the impedance control of robots with flexible elements. The concentration will be on series elastic actuators (SEA) and variable stiffness actuators (VSA). The cascade control combined with outer impedance loop is often proposed to these types of compliant actuators.

6.1 Constant impedance-joint robots

The general structure of flexible joint could consist of three components: the driver, the transmission system, and the elastic element which may be in series with the output link [20], please see Fig. 10.

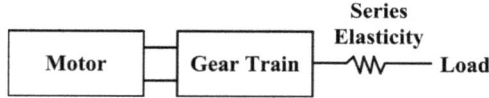

Fig. 10 Simplified schematic for series-elastic actuators (SEA) [86]

Accordingly, the actuator is called constant SEA or VSA based on the behaviour of the designed joint stiffness (constant stiffness or variable). In this category of actuators, the driver does not control the link directly but it will exchange energy with the transmission system which generates the flexible torque that actuates the link. In general, the dynamic modeling for flexible joints with constant or variable stiffness has the same mathematical formulation based on the assumptions of Spong [87]. The human-robot interaction may require actuators that can accurately and safely regulate the output torque. A SEA is a drive system in series with a compliant element (e.g., spring). Adding the elastic element in series with the driver and the output load could have the following characteristics [88]:

- It serves as an accurate torque source and as a low-cost torque sensor.
- The elastic element also serves as a compliant interface between the human and the robot, protecting the user and drive train from sudden shocks, and improving back drivability characteristics. In effect, the contact force can be regulated and controlled by the passive elements indirectly [89].
- The dynamic effects of backlash and friction could be small at the output of the elastic element.
- A drawback is the reduced large torque bandwidth due to motor saturation [90,91].

There are three important things that should be considered in designing flexible joints:

- The output flexible torque is important in the performance of interaction tasks; so considering the resulted flexibility of the elastic element as disturbance source could lead to loss of performance.
- If we have torque sensors, the design of the control law could be easy.
- The behavior of the flexible transmission could not be known completely for variable impedance actuators due to the inherent nonlinearity and the associated complexity.

In recent technology, flexible joints are integrated with robots to guarantee safe motion during contact phase or to attenuate the impact shock of unexpected forces [92,93]. The classical rigid body formulation for robots may be inadequate for motion in complicated tasks. The flexibility may exist due to the compliance of the gear transmission, belts and drive shafts, etc. However, adding flexibility to the joints could lead to some critical issues as follows: (i) the degree of freedom of the robotic system increases twice, (ii) the resulted system is not fully actuated due to the induced joint flexibility, (iii) the flexibility joint results in fast dynamics which may stimulate the vibration problems, and (iv) for motion in constrained space, a small deviation in joint position may lead to excessive contact force on the environment due to the coupling effect [94,95]. In general, the Lagrangian formulation for robots with flexible joints (e.g., simple harmonic drive, SEA, or even VSA) in constrained space can be expressed as

$$M(q)\ddot{q} + C(q,\dot{q})\dot{q} + g(q) = \tau_s + J^T\lambda \qquad (37a)$$

$$I_m\ddot{\theta} + B_m\dot{\theta} + \tau_s = u \qquad (37b)$$

with

$$\tau_s = K_s(\theta - q) \qquad (37c)$$

where I_m, B_m are the inertia and damping matrics on the motor side, θ is the motor angular position vector, and K_s is the spring stiffness matrix.

There are different techniques for dealing with control of flexible joints: decoupling control schemes [94,96-98], backstepping control [99], singular perturbation control [100], and adaptive control [101-104]. For discussion on limitations of these controllers, please refer to [105]. The control problem could be difficult when dealing with robots in constrained unknown space with joints flexibility [87]. The joint torque control is essential for vibration damping during the free space motion and soft and safe interaction control during the contact phase [105]. Modifications of eq. (37) to meet the requirements of inner torque control may demand the calculation of the fourth derivatives of the angular positions and the measurements of the derivatives of a torque sensor which could be quite noisy [95]. In effect, eq. (37) should be modified such that the full dynamics has output variables (q, τ_s) with the input control u. Albu-Schäffer and his colleagues [105,103] have concentrated on establishing high-performance impedance control strategy for torque controlled DLR robots avoiding the limitations of high derivatives measurements. DLR robots are provided with flexible joints, position sensors, and torques sensors. Combined impedance control and torque control are used for stabilization and control based on passivity theorem [105]. Second and higher order derivatives of the variables states are not required which give some preference on other techniques that use third derivatives of the variable states. The proposed control structure for the multi-loops control system is depicted in Fig. 11.

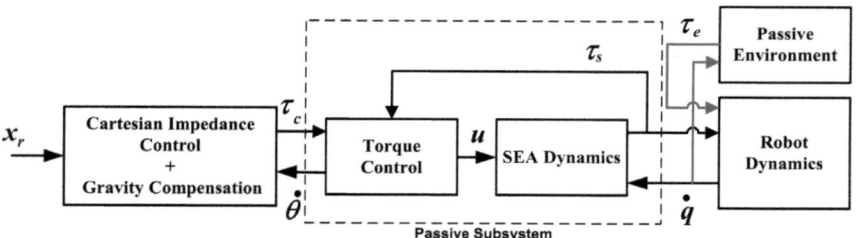

Fig. 11 The proposed control system for impedance control for flexible-joint DLR robot [105].

For robots with series elastic actuators, cascade control is often implemented with outer impedance loop and inner torque control as shown in Fig. 11. In effect, there are miscellaneous nested loops coupled with the impedance control; the details will be described below. Currently, the torque sensors are widely available [86]; however, using inner torque loop combined with other control loop is not always preferable and some challenges associated with the stability problem should be investigated. For example, the PI-based torque control could be difficult to be attuned or to provide high bandwidth if the load side having some damping. In addition, the control feedback could require torque sensor which is always noisy. Mosadeghzad et al. [86] have investigated the passivity and the implementation problems associated with different control loops (inner position control, inner torque control, and inner velocity control). It is interesting to mention the following observations:

- The inner loop with high bandwidth may not lead to a stable closed loop system when the outer impedance loop is closed for the range of values specified by the tasks requirements.
- The discrete impedance control system may require the lower bounds for the bandwidth of the inner control loop; however, in continuous time control system, larger values for the inner loop gains can be obtained to ensure stability and achieve high overall bandwidth.
- With the inner torque control loop, the model of the host robot should be known well to avoid instability. Whereas, the inner velocity loop can deal with the nonlinearity accompanied by the friction giving some robustness to the system.
- Unlike the velocity loop and torque loop, the position loop may give a stable closed loop system in free space. In effect, for a dynamic system having only velocity loop with zero reference velocity, the system attempts to stop but without tracking the desired position. Whereas, the system may keep moving with pure torque control.

Lagoda et al. [90] have designed an electric Series Elastic Actuated Joint for robotic gait rehabilitation training. The authors have designed multi-level control loops. The outer high-level control includes the impedance control that correlates the position and the interaction forces such that the deviation of the reference trajectory may lead to corrective forces. On the other hand, there are two nested control loops: torque control and velocity control that regulate the input control and the velocity respectively, please see Fig.12.

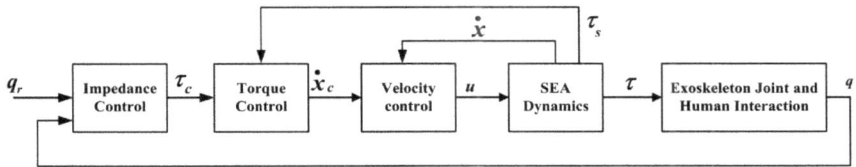

Fig. 12 The multi-level control of LOPES robotic system where the subscript refer to the commanded signals [90].

In effect, the same structure for impedance control of SEA has been proposed in [106].Tagliamonte et al. [106] have investigated the performance and stability conditions for three multi-level control loops: outer impedance loop having a virtual spring-damper system and an inner PI torque control generating a set-point reference for the innermost PI velocity loop. The authors proposed guidelines for tuning the controller gains and the possible ranges of virtual impedance parameters based on passivity theory generalizing the results of [106,107]

Zhao et al. [108] have proposed a critically-damped fourth order system gain selection criterion for a cascaded SEA control structure with inner torque and outer impedance feedback loops. Velocity filtering and feedback delays are taken into consideration for stability and impedance performance analysis. The proposed scheme is depicted in Fig. 13.

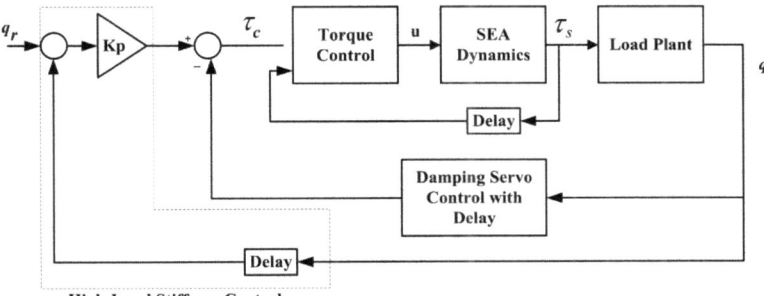

Fig. 13 The proposed cascade control with three control loops: high-level stiffness servo loop, inner embedded damping servo loop, and the innermost torque control [108]

The authors focused on maximizing the impedance range of SEA because most studies are concerned with low or near-zero impedance dynamics [109,110]. The authors showed that increasing the gains of the inner torque loop and decreasing the impedance gain may lead to instability problems. In addition, the SEA impedance converges to the virtual stiffness at a low frequency while it converges to spring stiffness at high frequencies. Yasuda et al. [111] have investigated the stability of impedance control of SEA with two motors. One motor is position controlled with high gear ratio reduction and high friction whereas the other side motor is torque controlled with low friction. Three case studies have been investigated: impedance control of position-controlled motor only, impedance control of SEA without a torque controlled motor, and impedance control of SEA with the torque-controlled motor. The stability of the actuator system has been improved with the torque-controlled motor. Valery et al. [112,107] have proposed cascade control with three nested loops: the outer impedance loop for regulating the relationship between the interaction force and the reference trajectory, an inner torque control, and the innermost velocity loop. The passivity conditions for the rendering of a pure spring are derived and the control gains are selected. In general, the following important points should be noted:

- Most work is concerned with using three nested control loops: innermost velocity loop, intermediate torque loop, and the outer impedance control that rendering the virtual impedance for safe and comfortable human-robot interactions, please see [90,106, 107,112,113].
- The powerful tool for tuning the gains of the multi-levelled loops and determining the ranges of the virtual impedance target could be based on passivity theory or the well-known stability strategies of linear control theory, Nyquist's stability, etc.
- For pure virtual spring impedance target, the maximum value of the virtual spring impedance is limited by the value of the physical stiffness of the SEA, the controller gains and the viscous friction (if exists).
- For zero impedance mode, the SEA is always passive if no integrators are used. For more details on more observations, the reader is referred to [113].
- Most work has concentrated on stability constraints of multi-level control for single joints using linear control theory. Extending the work for MIMO robots considering the nonlinearities, delay time problems, and the stability problems associated with multi-loops are not straightforward. Albu-Schäffer and his colleagues, however, have proposed a general framework to deal with impedance control of torque controlled actuator robots with MIMO [105,103]. Lin [95] has proposed a general framework for adaptive and robust control of flexible joint robots (with harmonic drive rather than SEA) considering the hybrid position/force control as an outer loop. See also the work of Li et al. [114] and the references therein for more details on dealing with human-robot interaction considering the MIMO adaptive control.

6.2 Variable impedance-joint robots

The added compliance of the new generation of actuators makes the robot intrinsically safe. The constant impedance actuators described previously may have some limitations associated with dealing with different tasks and motion frequencies; different tasks need variable stiffness (impedance) actions which could be lost in the SEAs. Therefore, a new generation of variable stiffness actuators has been built. In effect, robotic systems with controllable impedance (stiffness, damping, etc.) are capable for rejection of disturbances, storing the energy, and controlling the end-effector stiffness in contact space [115], please see [21,22] for more details on the VSA design, performance, and the classification of the VSA.

Accordingly, many efforts have been performed to design variable impedance actuators with different applications. However, in contrast to stiff actuators, a deviation of the desired position associated with the added compliance often accompanies the working principle of variable impedance actuators. Let us consider the simple variable stiffness actuator shown in Fig. 14 [139].

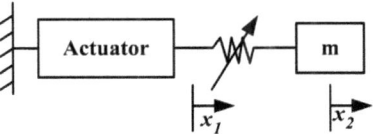

Fig. 14 A simple variable stiffness actuator connecting the conventional electrical actuator with the output load [139]

According to Fig. 14, there are two possible degrees of freedom that could be controlled: the equilibrium position displacement x_1 and the time-varying stiffness k(t).

In a similar way to variable impedance control described in section 5, the interaction force can be controlled by exploiting the passive variable stiffness elements. In effect, the idea of the variable stiffness actuators is to tune the system stiffness to adapt to any unexpected disturbance or interaction with the unknown environment [139]. In fact, most work concentrates on how to control the simultaneous position and stiffness control; there is a strong relationship between the system stiffness and the generalized torques (e.g., the input control). In light of above, some stiffness-based control schemes are considered below in order to take an insight into the interaction control of VIAs; there could be no clear outer impedance filter in most control strategies except with some references cited next. In general, the VIAs could be grouped into three categories [116,117]:

- Single actuated flexible transmission
- Antagonistic variable stiffness actuators
- Serial variable stiffness actuators

Please for more details on modeling and control of these groups, the reader is referred to [116].

Remark 7: Most work has been concentrating on position control of VAS with different schemes [118-120, 121-123], while there are only few works on interaction control of VSA-actuated robots with the known or unknown environment [123,118]. In effect, the feedback linearization [116] and the passivity-based approach [122,124,125] are the most powerful approaches to control the position/stiffness of the VSA. Note that this section is concerned with the cascade impedance control with inner torque control. It does not consider all the possible position control schemes for variable impedance actuators. Accordingly, we will attempt to cite literature work dealing with an outer impedance filter with inner torque control loop. In effect, combining the active impedance loop with the intrinsic impedance of the actuators could be limited, except the work of [117,126-128].

Based on neurophysiological investigations, the impedance regulation of human motion during interaction is basically performed at the muscles and joints levels and slightly at the end-effector level; this may contract the design philosophy of impedance control of robots which prefers the task space dynamics. Formica et al. [129] have proposed cascade control consisting of two loops: an outer force loop with tracking the desired forces based on a modification of the position, and an inner torque -based compliance control loop for replicating the human behavior. The stiffness matrix of the robot system is related to the input control with specified stiffness ranges. The proposed controller was applied for application of Robot-Mediated Motor Therapy (RMMT), please see Fig. 15 for a description of the proposed controller. Please for similar works, the reader is referred to [130-132].

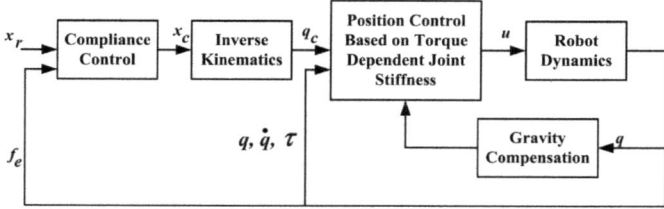

Fig. 15 The interaction force with torque-based compliance control for RMMT system [129]. As we see, the inner loop attempts to track the commanded joint positions based on torque-based stiffness

Best [115] has used cascade and successive control loops based on sliding mode control for controlling inflatable robots with pneumatically antagonistic actuators. As seen from Fig. 16, the control system consists of 5 subsystems: subsystem1 refers to the position control which generates the desired torque, subsystem 2 with the desired torque as input while the desired pressure is its output; this subsystem would control the torque and the system stiffness. Whereas the subsystem 3 is concerned with pressure control, subsystems 4 and 5 are specialized for states and stiffness observers. In effect, the subsystem 1 correlating the desired position, stiffness and the desired torques represents an impedance filter regulating the stiffness of the system associated to the interaction torques. In effect, this control scheme could show the important problems that should be considered in VSAs, e.g. the estimation of stiffness, the impedance filter that correlates the position, stiffness and the torque.

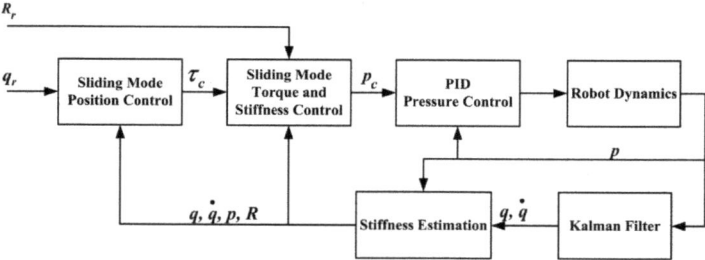

Fig. 16 Multi-level control loops of inflatable robotic systems with antagonistic pneumatic joints, where p refers to the pressure, PID is the proportional-integral-derivative control, and R denotes to the stiffness [115].

Liu et al. [128] have proposed a mechanical rotary impedance actuator with variable stiffness based on the controllable effective length of a mechanical bending bar. A cascade impedance control was proposed for controlling the motion and the stiffness of the target robot. It consists of three nested loops: an outer PD impedance loop, inner torque loop and the innermost velocity loop which is necessary for friction problems. The authors have used H_{inf} shaping controller for stabilization of the torque loop. The strategy is like that used in SEAs described previously, please see Fig. 17.

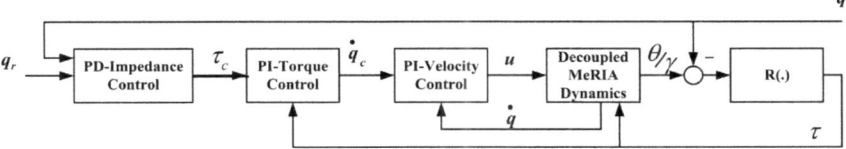

Fig. 17 The cascade control system for compliant actuators, where γ refers to the gear ratio, PD refers to the proportional-derivative controller, and PI is the integral-derivative controller. The stiffness controller design is neglected in this scheme. [128]

Wimböck et al. [126] have applied impedance control for variable stiffness mechanisms with nonlinear joint couplings. Two cascade loops are proposed: outer impedance loop with the desired position and stiffness references as inputs while the tendons forces and the variable lengths as outputs, and inner force control loop for tracking the desired tendon forces and lengths which are fed from the output of the outer impedance loop, please see Fig. 18.

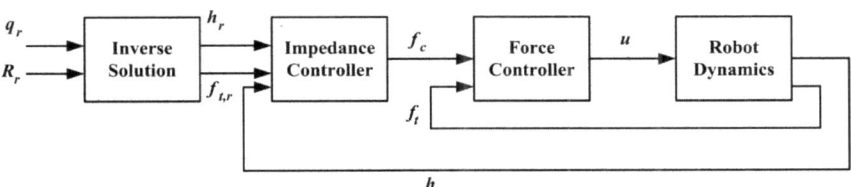

Fig. 18: Cascade impedance control with inner tendon force control loop, where the subscript t refers to the tendons, and h denotes to the tendon length changes [126].

Remark 6: Although section 6 is focusing on electro-mechanical actuators, a similar framework of cascade control can be used for controlling hydraulic actuators [133] as shown in Fig.19, and pneumatic actuators [115] described in Fig. 16.

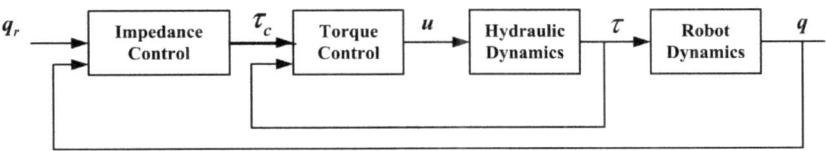

Fig. 19: Cascade impedance control with two control loops: an outer impedance loop for correlating the commanded torques and position, and an inner torque control which generates the input control u to the valve [133].

In summary, there some important points should be noted:

- For variable impedance actuators, the presence of two actuators for controlling the stiffness and the position simultaneously requires designing a special control law that take advantage of the actuators capabilities. The intended flexibility of the system should not be considered as vibration source as made in most conventional actuators [116].
- The complete behaviour of the transmission system is not simple to be known due to the accompanied nonlinearity and the design complexity, please see [133-137] for more details on estimation of system stiffness and damping.
- In general, there could be three control classifications for variable impedance actuated robots: (i) simultaneous control of position and stiffness control [138], (ii) impedance control with inner torque control [128,126], and (iii) bio-inspired control, e.g. time-based compliance control [129], co-activation control [130,131], etc.
- The assessment of the proposed control schemes, the performance, and the stability are not investigated deeply. The stability of impedance control associated with variable impedance actuators requires more studies. The impedance control, however, associated with inner torque control is the easiest control scheme to deal with constant and variable stiffness actuators. For more details on the control of variable stiffness actuators, the reader is referred to [124,125].

7. Conclusions

This paper attempts to systematically introduce features and limitations of categories of impedance control. Basically, the impedance control can be classified as force-based impedance control and position-based impedance control. The conventional impedance control schemes do not consider the force tracking problems in the outer impedance filter resulting in a deviation of the desired force references. Accordingly, then modification of the impedance filter to satisfy the force tracking problem is a motivating technique of imitation of human behavior. As aforementioned, one of the possible strategies for force tracking-based impedance control is changing the virtual stiffness. Therefore, a clear connection between the variable impedance control and the force tracking-based impedance control. On the other hand, changing the impedance parameters is not trivial. The investigation of stability problems of variable impedance control is rather mature and demands more work.

Impedance control of flexible joint actuated robots is a challenging problem. Controlling the robots with constant impedance joints could be easier than variable impedance joints. In variable impedance actuators, the stiffness is added variable output that should be carefully controlled. In general, an outer impedance filter integrated with torque control is a strong strategy for the solution of this category of transmission. The topic of impedance control variable impedance actuators is still early and demands more effort. The problem is that VSA is based on different design schemes and accordingly there is no unified control method that is suitable for all possible designs. Best [115], however, introduce interesting multi-level control that shows the basic control subsystems that should be considered with variable impedance actuated robots.

Reference

[1] T. Yoshikawa, Foundations of robotics: Analysis and control, MIT Press, 2003.

[2] L. Villani, Force control in robotics, John Baillieul, Tariq Samad (Ed.), Encyclopaedia of systems and control, Springer-Verlag, London 2015.

[3] Hayder F. N. Al-Shuka, Adaptive hybrid regressor and approximation control of robotic manipulators in constrained space, International Journal of Mechanical & Mechatronics Engineering, IJMME-IJENS, Vol:17, No:03, pp. 11-19, 2017.

[4] M.H. Raibert and J.J. Craig, Hybrid position/force control of manipulators. ASME J. Dynamic Meas. Control, Vol. 102, pp.126-133, 1981.

[5] N. Hogan, Impedance control: An approach to manipulation: Part I—Theory, ASME. J. Dyn. Sys., Meas., Control, Vol.107, No.1, pp.1-7, 1985.

[6] N. Hogan, Impedance control: An approach to manipulation: Part II—Implementation. ASME. J. Dyn. Sys., Meas., Control, Vol.107, No. 1, pp. 8-16, 1985.

[7] D. A. Lawrence, Impedance Control Stability Properties in Common Implementations, Impedance Control Stability Properties in Common Implementations, vol.2, pp. 1185 – 1190.

[8] R. Anderson, M. Spong, Hybrid impedance control of robotic manipulators, IEEE International Conference on Robotics and Automation, vol. 4, pp. 1073 – 1080, 1987.

[9] G. Liu and A. A. Goldenberg, Robust hybrid impedance control of robot manipulators via a tracking control method, Proceedings of the IEEE/RSJ/GI International Conference on Intelligent Robots and Systems, vol.3, pp. 1594–1601, 1994.

[10] Wen-Hong Zhu and J. De Schutter, Experimental verifications of virtual decomposition-based motion/force control, IEEE Transactions on Robotics and Automation, Vol. 18, No. 3, pp. 379-386, June 2002.

[11] H. Wang, K. H. Low and M. Y. Wang, Reference trajectory generation for force tracking impedance control by using neural network-based environment estimation, 2006 IEEE Conference on Robotics, Automation and Mechatronics, pp. 1-6, 2006.

[12] W. Xu, C. Cai and Y. Zou, Neural-network-based robot time-varying force control with uncertain manipulator-environment system, Transactions of the Institute of Measurement and Control, Vol 36, No. 8, pp. 999–1009, 2014.

[13] H. Wang, K. H. Low and M. Y. Wang, Reference trajectory generation for force tracking impedance control by using neural network based environment estimation, IEEE Conference on Robotics, Automation and Mechatronics, pp. 1–6, 2006.

[14] L. Roveda et al. Optimal impedance force-tracking control design with impact formulation for interaction tasks, IEEE Robotics and Automation Letters, Vol. 1, No. 1, pp. 130-136, 2016.

[15] S. Jung and T. C. Hsia, Force tracking impedance control of robot manipulators for environment with damping, IECON 2007-33rd Annual Conference of the IEEE Industrial Electronics Society, pp. 2742–2747, 2007.

[16] S. Jung and T. C. Hsia, Stability and convergence analysis of robust adaptive force tracking impedance control of robot manipulators, Proceedings IEEE/RSJ International Conference on Intelligent Robots and Systems, Vol. 2, pp. 635–640, 1999.

[17] S. Jung and T. C. Hsia, Reference compensation technique of neural force tracking impedance control for robot manipulators, The 8th World Congress on Intelligent Control and Automation, pp. 650 – 655, 2010.

[18] B. Komati, C. Clevy and P. Lutz, Force tracking impedance control with unknown environment at the microscale, IEEE International Conference on Robotics and Automation (ICRA), pp. 5203–5208, 2014,

[19] K. Kronander and A. Billard, Stability considerations for variable impedance control, IEEE Transactions on Robotics, Vol. 32, No. 5, pp. 1298–1305, 2016.

[20] F. Petit, Analysis and control of variable stiffness robot, Doctor of Sciences of Eth Zurich, 2014.

[21] B. Vanderborght et al., Variable impedance actuators: A review, Robotics and Autonomous Systems, Vol. 61, No.12, pp. 1601–1614, 2013.

[22] R. V. Ham, T. G. Sugar, B. Vanderborght, K. W. Hollander, and D. Lefeber, Compliant actuator designs, IEEE Robotics & Automation Magazine, Vol. 16, No. 3, pp. 81–94, 2009.

[23] M. Vukobratovic, D. Surdilovic, Y. Ekalo and D. Katic, Dynamics and robust control of robot-environment interaction, World Scientific Publishing Company; 1 Edition (March 6, 2009).

[24] J.K. Salisbury, Active stiffness control of a manipulator in Cartesian coordinates, 19th IEEE Conf. Decis. Contr. Albuquerque, pp. 95–100, 1.980

[25] L. Villani and J. De Schutter, Force control in Springer Handbook of Robotics, Editors: Bruno Siciliano, Oussama Khatib, pp 161-185, 2008.

[26] Y. Oh, W. K. Chung, Y. Youm and M. Kim, Hybrid impedance control of redundant manipulators: an approach to decouple task space and null space motions, Intelligent Automation & Soft Computing, Vol.5, No.2, pp.149-164, 1999.

[27] D. E. Whitney, Force feedback control of manipulator fine motions, ASME J. Dyn. Syst. Meas. Contr. Vol.99, pp. 91–97, 1977.

[28] O. Khatib, A Unified approach for motion and force control of robot manipulators: the operational space formulation, IEEE Journal on Robotics and Automation, Vol. 3, No.1, pp. 43–53, 1987.

[29] M. Ali, Impedance control of redundant manipulators, PhD Dissertation, Tampere University of Technology, Finland, 2011.

[30] C. Ott, Cartesian impedance control of redundant and flexible-joint robots, Springer Tracts in Advanced Robotics, 2008.

[31] A. Dietrich, Whole-body impedance control of wheeled humanoid robots, Springer Tracts in Advanced Robotics, 2016.

[32] A.-C. Huang, and M. -C. Chien, Adaptive Control of Robot Manipulators: A Unified Regressor-free Approach, World Scientific Publishing Company, 2010.

[33] O. Khatib, "Motion/Force Redundancy of Manipulators," JAPAN-USA. Symp. on Flexible Automation, pp. 337-342, 1990

[34] O. Khatib, "Inertial Properties in Robotic Manipulation: An Object-Level Framework," Int. J. of Robotics Research, Vol. 14, No. 1, pp. 19-36, 1995

[35] C. Pholsiri, D. Rabindran, M. Pryor and C. Kapoor, Extended Generalized Impedance Control for Redundant Manipulators, The 42nd IEEE International Conference on Decision and Control (IEEE Cat. No.03CH37475), Vol. 4, pp. 3331 – 3336, 2003.

[36] C. Ott, , A. Dietrich, A. Albu-Schäffer, Prioritized multi-task compliance control of redundant manipulators, Automatica, Vol. 53, pp. 416–423, 2015.

[37] N. Oda, H. Ohta, T. Murakami and K. Ohnishi, A robust impedance control strategy for redundant manipulator, Proceedings of the 1995 IEEE IECON 21st International Conference on Industrial Electronics, Control, and Instrumentation, Vol. 2, 1254 – 1259, 1995.

[38] H. Seraji and R. Colbaugh, Force tracking in impedance control, IEEE International Conference of Robotics and Automation, Vol. 2, pp. 499-506, 1993.

[39] A. De Luca, Modeling and Control of Robots with Elastic Joints, lectures Notes, http://www.diag.uniroma1.it/~deluca/flexiblejoints.html.

[40] A. M. Khan, D.-W. Won, M. A. Ali, J. Han, K. Shan, and C. Han, Adaptive impedance control for upper limb assist exoskeleton, IEEE International Conference on Robotics and Automation, pp. 4359 – 4366, 2015.

[41] C. Ott, R. Mukherjee and Y. Nakamura, A hybrid system framework for unified impedance and admittance control, Journal of Intelligent & Robotic Systems, , Vol. 78, No. 3, pp 359–375, 2015.

[42] Z. M. Sotirov and R. G. Botev, A model reference approach to adaptive impedance control of robot manipulators, IEEE/RSJ International Conference on Intelligent Robots and Systems IROS, Vol. 2, pp. 727–733, 1993.

[43] M. Sharifi, S. Behzadipour and G. Vossoughi, Nonlinear model reference adaptive impedance control for human–robot interactions, Control Engineering Practice, Vol 32, pp. 9-27, 2014.

31

[44] R. Kamnik, D. Matko and T. Bajd, Application of model reference adaptive control to industrial robot impedance control, Journal of Intelligent and Robotic Systems, Vol. 22, No.2, pp.153-163 1998.

[45] R. Collbaugh and K. Glass, Adaptive compliant motion control of manipulators without velocity measurements, Journal of Robotic Systems, Vol.14, No.7, pp.513–527, 1997.

[46] W. -S. Lu and Q. -H Meng, Impedance control with adaptation for robotic manipulations, IEEE Transactions on Robotics and Automation, Vol.7, No.3, pp. 408-415, 1991.

[47] B. Alquadi, H. Modares, I. Ranatunga, S. M. Tousif, F. L. Lewis, and D. O. Popa, Model reference adaptive impedance control for physical human-robot interaction, Control Theory and Technology, Vol.14, No.1, pp. 68-82, 2016.

[48] R. Kelly, R. Carelli, M. Amestegui, and R. Ortega, On adaptive impedance control of robot manipulators, IEEE International Conference on Robotics and Automation, Vol. 1, pp. 571-577, 1989.

[49] K. Lee and M. Buss, Force tracking impedance control with variable target stiffness, The International Federation of Automatic Control IFAC, Seoul, Korea, Vol.41, No.2, pp.6751-6756, 2008.

[50] W. Xu, C. Cai, M. Yin and Y. Zou, Time-varying force tracking in impedance control. 51st IEEE Conference on Decision and Control, Maui, Hawaii, USA, pp. 344-349, 2012.

[51] W. Xu, C. Cai, Y. Zou, Neural-network-based robot time-varying force control with uncertain manipulator–environment system, Transactions of the Institute of Measurement and Control, Vol 36, No.8, pp. 999–1009, 2014.

[52] H. Huang, D. Sun, J.K. Mills, S. H. Cheng, Integrated vision and force control in suspended cell injection system: Towards automatic batch biomanipulation, IEEE international Conference on Robotics and Automation, pp. 3413–3418, 2008.

[53] Y. Xie, D. Sun, C. Liu, S.H. Cheng, A force control approach to a robot-assisted cell microinjection system, The International Journal of Robotics Research, Vol.29, No.9, pp.1222–1232, 2010.

[54] C.T. Freeman, E. Rogers, A.-M. Hughes A, J.H. Burridge and K.L. Meadmore, Iterative learning control in health care: Electrical stimulation and robotic assisted upper-limb stroke rehabilitation. IEEE Control Systems Magazine, Vol. 32, No.1, pp.18–53, 2012.

[55] B. Varkuti et al., Resting state changes in functional connectivity correlate with movement recovery for BCI and robot-assisted upper-extremity training after stroke. Neurorehabilitation and Neural Repair27(1): 53–62, 2012.

[56] T. Kim, H. S. Kim and J. Kim. Position-based impedance control for force tracking of a wall-cleaning unit. International Journal of Precision Engineering and Manufacturing, Vol. 17, No. 3, pp. 323-329, 2016.

[57] R. A. Volpe, Real and artificial forces in the control of manipulators: theory and experiments, Ph.D. Dissertation, Carnegie Mellon University, USA, 1990.

[58] O. Khatib and J, Burdick, Motion and force control of robot manipulators. Proc. IEEE Int. Conf. Robot. and Automat. Los Alamitos, CA, pp. 1381-1386, 1986.

[59] F. Ficuciello, L. Villani and B. Siciliano, Variable impedance control of redundant manipulators for intuitive human–robot physical interaction, IEEE Transactions on Robotics, Vol. 31, No. 4, pp. 850-863, 2015.

[60] M. N. Oğuztöreli, and R. B. Stein,The effect of variable mechanical impedance on the control of antagonistic muscles,Biological Cybernetics, Vol. 66, No.2, pp 87–93, 1991.

[61] W. A. Farahat, H. Herr, Optimal workloop energetics of muscle-actuated systems: an impedance matching view, PLoS Computational Biology, Vol. 6, No. 6, pp. 1-11, 2010.

[62] H. Tomori, S. Nagai, T. Majima and T. Nakamura, Variable impedance control with an artificial muscle manipulator using instantaneous force and mr brake, IEEE/RSJ International Conference on Intelligent Robots and Systems, pp. 5396 – 5403, 2013.

[63] D. Heck, D. Kostic, A. Denasi, and H. Nijmeije, Internal and external force-based impedance control for cooperative manipulation, In European Control Conference (ECC), pp. 2299-2304., 2013.

[64] F. Ferraguti, C. Secchi and C. Fantuzzi, A Tank-based approach to impedance control with variable stiffness, IEEE International Conference on Robotics and Automation, pp. 4948–4953, 2013.

[65] J. H. Park and H. C. Cho, Impedance control with varying stiffness for parallel-link manipulators, Proceedings of the 1998 American Control Conference, Vol. 1, pp. 478–482,1998.

[66] G. Ganesh, N. Jarrassé, S. Haddadin, A. Albu-Schaeffer and E. Burdet, A versatile biomimetic controller for contact tooling and haptic exploration, 2012 IEEE International Conference on Robotics and Automation, pp. 3329–3334, 2012.

[67] B. Yuan B, M. Sekine, J. Gonzalez, J.G. Tames, W.Yu, Variable Impedance Control Based on Impedance Estimation Model with EMG Signals during Extension and Flexion Tasks for a Lower Limb Rehabilitation Robotic System. J Nov Physiother, Vol.5, No.3, 2013, 178. doi:10.4172/2165-7025.1000178.

[68] G. A.- Ollinger, J. E. Colgate, M. A. Peshkin and A. Goswami, Active-impedance control of a lower-limb assistive exoskeleton, Proceedings of the 2007 IEEE 10th International Conference on Rehabilitation Robotics, June 12-15, Noordwijk, The Netherlands, pp. 188-192, 2007.

[69] T. Tsumugiwa, R. Yokogawa and K. Hara ,Variable impedance control with virtual stiffness for human-robot cooperative task(human-robot cooperative peg-in-hole task), Proceedings of the 41st SICE Annual Conference. SICE 2002., Vol. 4, pp. 2329 – 2334, 2002.

[70] B. Berret, G. Sandini and F. Nori, Design principles for muscle-like variable impedance actuators with noise rejection property via co-contraction, IEEE-RAS International Conference onHumanoid Robots (Humanoids), pp. 222-227, 2012.

[71] R. Schiavi, G. Grioli, S. Sen, and A. Bicchi, VSA-II: a novel prototype of variable stiffness actuator for safe and performing robots interacting with humans, IEEE International Conference on Robotics and Automation, pp. 2171–2176, 2008.

[72] A. Jafari, N. G. Tsagarakis, B. VanderborghtD and. G. Caldwell, A novel actuator with adjustable stiffness (AwAS), IEEE/RSJ International Conference on Intelligent Robots and Systems, pp. 4201-4206, 2010.

[73] S. Wolf, and G. Hirzinger, A new variable stiffness design: Matching requirements of the next robot generation, 2008 IEEE International Conference on Robotics and Automation, pp. 1741 – 1746, 2008.

[74] E. Burdet, R. Osu, D. Franklin, T. E. Milner, and M. Kawato, The central nervous system stabilizes unstable dynamics by learn-ing optimal impedance, Nature, Vol. 414, No. 6862, pp. 446–449, 2001.

[75] M. Howard, D. Braun, and S. Vijayakumar, "Transferring hu-man impedance behavior to heterogeneous variable impedance ac-tuators," IEEE Trans. Robot., vol. 29, no. 4, pp. 847–862, Aug.2013

[76] A. Ajoudani, N. Tsagarakis, and A. Bicchi, Tele-impedance: Teleoperation with impedance regulation using a body-machine interface, Int. J. Robot. Res., Vol. 31, No. 13, pp. 1642–1656, 2012.

[77] H. He, M. Luo, and Q. Zhang, Dual impedance control with variable object stiffness for the dual-arm cooperative manipulators, IEEE Asia-Pacific Conference on Intelligent Robot Systems (ACIRS), pp. 102-108, 2016.

[78] R.J. Butler, H.P. Crowell and I.M. Davis, Lower extremity stiffness: implications for performance and injury, Clin Biomech (Bristol, Avon), Vol.18, No.6, pp.511-7, 2003.

[79] S. Rapoport, J. Mizrahi, E. Kimmel, O. Verbitsky and E. Isakov, Constant and variable stiffness and damping of the leg joints in human hopping, J Biomech Eng, Vol.125, No.4, pp.507-514, 2003.

[80] M. Plocharski and P. Plocharski, Ankle joint stiffness during phases of human walking, MSc Dissertation, Department of Health Science & Technology, Aalborg University, 2013.

[81] F. Ficuciello, L. Villani, B. Siciliano, Impedance control of redundant manipulators for safe human-robot collaboration, Acta Polytechnica Hungarica, Vol.13, No.1, pp.223-238, 2016.

[82] F. Ferraguti et al. An energy tank-based interactive control architecture for autonomous and teleoperated robotic surgery, IEEE Transaction on Robotics, Vol. 31, No. 5, pp. 1073-1088, 2015.

[83] V. VanDoren, Fundamentals of cascade control, http://www.controleng.com/single-article/fundamentals-of-cascade-control/bcedad6518aec409f583ba6bc9b72854.html

[84] T. Liu and F. Gao, Industrial, Process identification and control design: step-test and relay-experiment-based methods, Advances in Industrial Control, Springer-Verlag, Chapter 9 Cascade control system, 2014.

[85] R. Ghorbani, On controllable stiffness bipedal walking, PhD Dissertation. Department of Mechanical and Manufacturing Engineering, The University of Manitoba, Canada, 2008.

[86] M. Mosadeghzad, G. A. Medrano-Cerda, J. A. Saglia, N. G. Tsagarakis and D. G. Caldwell, Comparison of various active impedance control approaches, modeling, implementation, passivity, stability and trade-offs, 2012 IEEE/ASME International Conference on Advanced Intelligent Mechatronics (AIM), pp. 342 – 348, 2012.

[87] M. W. Spong, Modeling and control of elastic joint robots, ASME J. of Dynamic Systems, Measurement, and Control, Vol.109, No.4, pp.310-319, 1987.

[88] G. Pratt and M. Williamson, Series elastic actuators, IEEE/RSJ International Conference on Intelligent Robots and Systems 95. 'Human Robot Interaction and Cooperative Robots', Proceedings, Vol. 1, pp. 399–406, 1995.

[89] B.-S. Kim and J.-B.Song, Object grasping using a 1 dof variable stiffness gripper actuated by a hybrid variable stiffness actuator, 2011 IEEE International Conference on Robotics and Automation, Shanghai, China, pp. 4620-4625, 2011.

[90] C. Lagoda, A. C. Schouten, A. H. A. Stienen, E. G. Hekman and H. van der Kooij, Design of an electric Series Elastic Actuated Joint for robotic gait rehabilitation training, The 3rd IEEE RAS & EMBS International Conference on Biomedical Robotics and Biomechatronics, pp. 21-26, 2010.

[91] J. E. Pratt, B. Krupp, and C. Morse. Series elastic actuators for high fidelity force control. Industrial Robot: An international robot, Vol.29, No.3, pp.234–241, 2002.

[92] M. Zinn, O. Khatib and B. Roth, A new actuation approach for human frierdly robot design. International Journal of Robotics Research, Vol.23, pp.279–398, 2004.

[93] A. Bicchi, G. Tonietti, M. Bavaro, and M. Piccigallo, Variable stiffness actuators for fast and safe motion control, In: Dario P., Chatila R. (eds) Robotics Research. The Eleventh International Symposium. Springer Tracts in Advanced Robotics, Vol 15. Springer, Berlin, Heidelberg, 2005.

[94] T. Lin and A.A. Goldenberg, Robust adaptive control of flexible joint robots with joint torque feedback. IEEE International Conference on Robotics and Automation, pp.1229–1234, 1995.

[95] T. Lin, Adaptive and robust control of flexible joint robots with joint torque feedback, PhD dissertation, Department of Electrical and Computer Engineering University of Toronto, Canada, 1995.

[96] M. Spong, Modeling and control of elastic joint robots. ASME Journal of Dynamic Systems, Measurement, and Control, Vol.109, pp.310–319, 1987.

[97] B. Brogliato, R. Ortega and R. Lozano, Global tracking controllers for flexible-joint manipulators: a comparative study, Automatica, Vol.31, No.7, pp.941–956, 1995.

[98] A. De Luca and P. Lucibello, A general algorithm for dynamic feedback linearization of robots with elastic joints, IEEE International Conference on Robotics and Automation. Leuven, Belgium, Vol.1, pp.504-510, 1998.

[99] C.-M. Ou, J.-S. Lin, Nonlinear adaptive backsteppingcontrol design of flexible-joint robotic manipulators, 2011 8th Asian Control Conference (ASCC), pp. 1352 – 1357, 2011.

[100] M. W. Spong, J. Y. Hung, S. A. Bortoff and F. Ghorbel, A comparison of feedback linearization and singular perturbation techniques for the control of flexible joint robots, American Control Conference, pp. 25–30, 1989.

34

[101]   M. Spong, Adaptive control of flexible joint manipulators, Systems and Control Letters, Vol.13, pp. 15–21, 1989.

[102]   S. Nicosia and P. Tomei, A method to design adaptive controllers for flexible joint robots, IEEE International Conference on Robotics and Automation, pp. 701–706, 1992.

[103]   C. Ott, A. Albu-Schäffer and G. Hirzinger, Comparison of adaptive and nonadaptive tracking control laws for a flexible joint manipulator, IEEE International Conference on Intelligent Robotic Systems, pp. 2018–2024, 2002.

[104]   W.-H. Zhu and J. De Schutter, Adaptive control of mixed rigid/flexible joint robot manipulators based on virtual decomposition, IEEE Transactions on Robotics and Automa-tion, Vol.15, No.2, pp. 310–317, 1999.

[105]   A. Albu-Schäffer, C. Ott, G. Hirzinger, A unified passivity based control framework for position, torque and impedance control of flexible joint robots, The international journal of robotics research, Vol.26, No.1, pp.23-39, 2007.

[106]   N.L. Tagliamonte, D. Accoto, E. Guglielmelli, Rendering viscoelasticity with series elastic actuators using cascade control, IEEE International Conference on Robotics and Automation (ICRA), pp. 2424–2429, 2014.

[107]   H. Vallery, R. Ekkelenkamp, H. van der Kooij and M. Buss, Passive and accurate torque control of series elastic actuators. In:IEEE/RSJ International Conference on Intelligent Robots and Systems, San Diego, CA, pp.3534–3538, 2007.

[108]   Y. Zhao, N. Paine and L. Sentis, Feedback parameter selection for impedance control of series elastic actuators, IEEE-RAS International Conference on Humanoid Robots, pp. 999–1006, 2014.

[109]   G. A. Pratt, P. Willisson, C. Bolton, and A. Hofman, Late motor processing in low-impedance robots: Impedance control of series-elastic actuators, IEEE American Control Conference, Vol. 4, pp. 3245–3251, 2004.

[110]   K. Kong, J. Bae, and M. Tomizuka, Control of rotary series elastic actuator for ideal force-mode actuation in human–robot interaction applications, IEEE/ASME Transactions on Mechatronics, Vol. 14, No. 1, pp. 105–118, 2009.

[111]   K. Yasuda, R. Ikeura, S. Hayakawa and H. Sawai, Stability of impedance control of a series elastic actuator with a torque controlled actuator for improving the system performance, IEEE International Conference on Robotics and Biomimetics (ROBIO), p. 2163 – 2168, 2015.

[112]   H. Vallery et al., Compliant actuation of rehabilitation robots: benefits and limitations of series elastic actuators, IEEE Robot Autom Mag, Vol.15, No.3, pp 60–69, 2008.

[113]   N. L. Tagliamonte, D. Accoto, Passivity constraints for the impedance control of series elastic actuators, Proceedings of the Institution of Mechanical Engineers, Part I: Journal of Systems and Control Engineering, Vol 228, No. 3, pp. 138 – 153, 2013.

[114]   X. Li, Yo. Pan, G. Chen, H. Yu, Adaptive human–robot interaction control for robots driven by series elastic actuators, IEEE Transactions on Robotics, Vol. 33, No. 1, pp. 169 – 182, 2017.

[115]   C. M. Best, Position and stiffness control of inflatable robotic links using rotary pneumatic actuation, MSc Dissertation, Brigham Young University, USA, 2016.

[116]   F. Flacco, Modeling and control of robots with compliant actuation, Dipartimento di Ingegneria Informatica, Automatica e Gestionale, SAPIENZA Universit´a di Roma, PhD Dissertation, 2012.

[117]   F. P. Petit, Analysis and control of variable stiffness robots, PhD Dissertation, Sciences of ETH Zurich, 2014.

[118]   G. Tonietti, R. Schiavi, and A. Bicchi, Design and control of a variable stiffness actuator for safe and fast physical human/robot interaction, IEEE International Conference on Robotics and Automation, pp. 526–531, 2005.

[119]   G. Palli, C. Melchiorri, T. Wimbock, M. Grebenstein and G.Hirzinger, Feedback linearization and simultaneous stiffness-position control of robots with antagonistic actuated joints, IEEE International Conference on Robotics and Automation, pp. 4367–4372, 2007.

[120]   J. Nakanishi, K. Rawlik and S. Vijayakumar, Stiffness and temporal optimization in periodic movements: An optimal control approach, IEEE/RSJ International Conference on Intelligent Robots and Systems (IROS), pp. 718–724, 2011.

[121]   I. Sardellitti, G. Medrano-Cerda, N. G. Tsagarakis, A. Jafari and D. G. Caldwell, A position and stiffness control strategy for variable stiffness actuators, IEEE International Conference on Robotics and Automation, pp. 2785–2791, 2012.

[122]   G. Buondonno and A. De Luca, Efficient computation of inverse dynamics and feedback linearization for vsa-based robots, IEEE Robotics and Automation Letters , Vol.1, No.2, pp. 908-917, 2016.

[123]   T.R. Kerkhof, Control strategy for variable stiffness actuators in bilateral teleimpedance interaction tasks, EEMCS / Electrical Engineering, MSc Report, 2013.

[124]   P. Erler, P. Beckerle, B. Strah and S. Rinderknecht, Experimental comparison of nonlinear motion control methods for a variable stiffness actuator, 5th IEEE RAS/EMBS International Conference on Biomedical Robotics and Biomechatronics, pp. 1045–1050, 2014.

[125]   M. Lendermann, B. R. P. Singh, F. Stuhlenmiller, P. Beckerle, S. Rinderknecht and P. V. Manivannan, Comparison of passivity based impedance controllers without torque-feedback for variable stiffness actuators, IEEE International Conference on Advanced Intelligent Mechatronics (AIM), pp. 1126–1131, 2015.

[126]   T. Wimböck, C. Ott, A. Albu-Schaffer, A. Kugi and G. Hirzinger, Impedance control for variable stiffness mechanisms with nonlinear joint coupling, IEEE/RSJ International Conference on Intelligent Robots and Systems, pp. 3796–3803, 2008.

[127]   F. Petit, A. Albu-Schäffer, Cartesian impedance control for a variable stiffness robot arm, IEEE/RSJ International Conference on Intelligent Robots and Systems, pp. 4180 – 4186, 2011.

[128]   L. Liu, S. Leonhardt and B.J.E. Misgeld, Design and control of a mechanical rotary variable impedance actuator, Mechatronics, Vol. 39, pp 226–236, 2016.

[129]   D. Formica, L. Zollo and E. Guglielmelli, Torque-dependent compliance control in the joint space for robot-mediated motor therapy, Journal of Dynamic Systems Measurement and Control, Vol.128, No.1, pp. 152-185, 2006.

[130]   L. Zollo, B. Siciliano, E. Guglielmelli, and P. Dario, A bio-inspired approach for regulating visco-elastic properties of a robot arm, IEEE International Conference on Robotics and Automation, Taipei, Taiwan, pp. 14–19, 2003.

[131]   L. Zollo, L. Dipietro, B. Siciliano, E. Guglielmellim and P. Dario, A bio-inspired approach for regulating and measuring visco-elastic properties of a robot arm, J. Rob. Syst., Vol.22, pp. 397–419, 2005.

[132]   H. Gomi and R. Osu, R., Task-dependent viscoelasticity of human multi-joint arm and its spatial characteristics for interaction with environments, J. Neurosci., Vol.18, pp. 8965–8978, 1989.

[133]   T. Boaventura, J. Buchli, C. Semini and D. G. Caldwell, Model-based hydraulic impedance control for dynamic robots, IEEE Transactions on Robotics, Vol. 31, No. 6, pp. 1324–1336, 2015.

[134]   F. Flacco, A. De Luca, I. Sardellitti, N. G. Tsagarakis, On-line estimation of variable stiffness in flexible robot joints, The International Journal of Robotics Research, Vol 31, No.13, pp. 1556–1577, 2012.

[135]   A. Serio, G. Grioli, I. Sardellitti, N. G. Tsagarakis and A. Bicchi, A decoupled impedance observer for a variable stiffness robot, IEEE International Conference on Robotics and Automation, pp. 5548 – 5553, 2011.

[136]   T. Ménard, G. Grioli and A. Bicchi, A stiffness estimator for agonistic-antagonistic variable-stiffness-actuator devices, IEEE Transactions on Robotics, 2014, Vol. 30, No. 5,pp. 1269–1278, 2014.

[137]   N. Kashiri, G. A. Medrano-Cerda, N. G. Tsagarakis, M. Laffranchi and Darwin Caldwell, Damping control of variable damping compliant actuators, IEEE International Conference on Robotics and Automation (ICRA), pp. 850–856, 2015.

[138]   A. Albu-Schäffer, S. Wolf, O. Eiberger, S. Haddadin, F. Petit and M. Chalon, Dynamic modelling and control of variable stiffness actuators, IEEE International Conference on Robotics and Automation, pp. 2155–2162, 2010.

[139]   L. Visser, Variable stiffness actuator: Modeling, control, and application to compliant bipedal walking, PhD Dissertation, Universiteit Twente, The Netherlands, 2013.